Oxford **Mathematics**
Primary Years Programme

1

Contents

OXFORD
UNIVERSITY PRESS
AUSTRALIA & NEW ZEALAND

Practice

1 Complete the number lines.

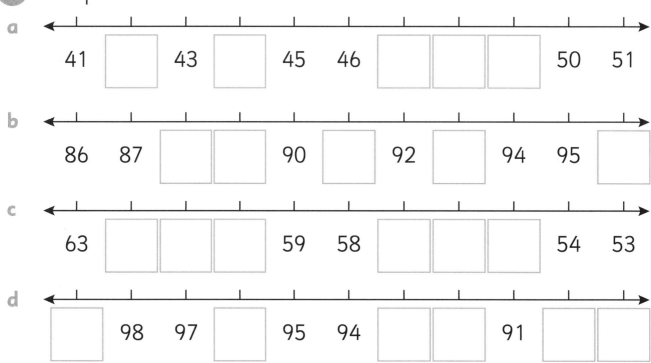

a

41 ☐ 43 ☐ 45 46 ☐ ☐ ☐ 50 51

b

86 87 ☐ ☐ 90 ☐ 92 ☐ 94 95 ☐

c

63 ☐ ☐ ☐ 59 58 ☐ ☐ ☐ 54 53

d

☐ 98 97 ☐ 95 94 ☐ ☐ 91 ☐ ☐

2 Write these numbers in the correct places on the hundred chart.

19　82　45　67　33　6　90　71

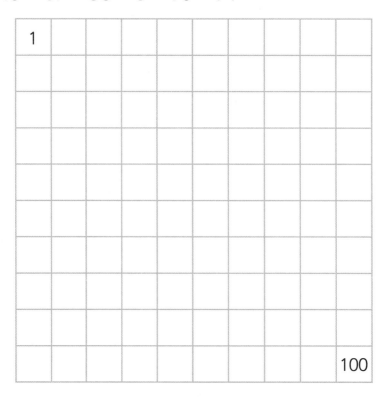

OXFORD UNIVERSITY PRESS

1 Tom and Lexi play 'First to 30'. Help Lexi work out how many counters Tom has on his ten-frames.

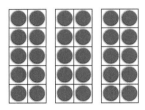

In 'First to 30', the first person to make 30 by filling three ten-frames wins!

a 6 spaces left = ☐ counters on the ten-frames.

b 11 spaces left = ☐ counters on the ten-frames.

c 18 spaces left = ☐ counters on the ten-frames.

d 9 spaces left = ☐ counters on the ten-frames.

e 23 spaces left = ☐ counters on the ten-frames.

2 Ana and Stella live on the same street. The difference between their house numbers is 14. What could their house numbers be?

Ana's house number	Stella's house number

You might like to use a number line to help you.

1 Lexi rolls a dice six times and makes exactly 30. What numbers could she have rolled?

a _____ + _____ + _____ + _____ + _____ + _____ = 30

b _____ + _____ + _____ + _____ + _____ + _____ = 30

c _____ + _____ + _____ + _____ + _____ + _____ = 30

d _____ + _____ + _____ + _____ + _____ + _____ = 30

e _____ + _____ + _____ + _____ + _____ + _____ = 30

2 Guess the numbers.

a I am an even number. The difference between the two digits in my number is 2. What number might I be?

b I am an odd number. The difference between my two digits is 3. What number might I be?

OXFORD UNIVERSITY PRESS

Practice

1 Write these house numbers in words.

 23   17   41   66

_____   _____   _____   _____

2 Match each person's age to the correct birthday cake.

27

12

50

18

35

twenty-seven

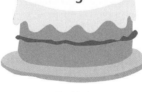
thirty-five

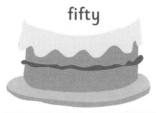
fifty

Challenge

1 Roll two dice and use the numbers to make a 2-digit number. Write the number in numerals and words. Draw it in pictures e.g. cubes. Repeat this two more times.

Number (numerals)	Number (words)	Number (pictures)

2 What other 2-digit numbers could you have made from the numbers you rolled in question 1?

I rolled a 3 and a 5. I could have made 35 or 53!

My number	Other number I could have made

3 What number is shown by **x** on the number lines? Write it in numerals and words.

Number line	Number (numerals)	Number (words)
18 19 20 X 22 23 24		
34 35 36 37 X 39 40		
27 X 25 24 23 22 21		
43 42 X 40 39 38 37		
89 X 91 92 93 94 95		

OXFORD UNIVERSITY PRESS

1 Jess saw beetles crawling up a tree. Write how many beetles she might have seen. Each beetle has six legs. Work out how many legs.

Number of beetles	Number of legs (numerals)	Number of legs (words)
e.g. 2	12	twelve

2 There are seven 2-digit numbers whose digits when added together equal 7. Write each of them in numerals and in words.

e.g. 16   sixteen

_____   _____

_____   _____

_____   _____

_____   _____

_____   _____

_____   _____

> *1 and 6 is 7. What other numbers can be combined to make 7?*

Practice

1 Put these numbers in the correct places on the number line.

a 54   b 39   c 43

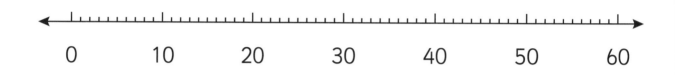

d 66   e 51   f 79

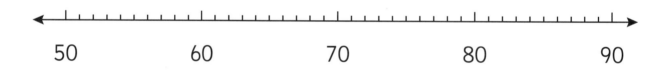

2

a Roll two 10-sided (0–9) dice. Combine the numbers to make a 2-digit number. Record the numbers below. Repeat this five more times.

_____ _____ _____ _____ _____ _____

b Order your six numbers from smallest to largest.

Smallest   Largest

_____ _____ _____ _____ _____ _____

OXFORD UNIVERSITY PRESS

Challenge

1 Roll two 10-sided dice. Combine the numbers to make a 2-digit number. Write them below. Repeat this seven more times.

_____   _____   _____   _____   _____   _____   _____   _____

Use your numbers to answer the following questions.

a Which number is the largest? _____

b Which number is the smallest? _____

c Which numbers are odd? _____

d Which numbers are even? _____

e Which numbers are less than 50? _____

f Which numbers are more than 50? _____

2

a How many different 2-digit numbers can you make from these numbers?

2   7   4   5

```



```

b Order the numbers you made from largest to smallest.

_____

_____

Mastery

1 These are the basketball rankings for the top 7 teams in the league.

Team	Red	Blue	Green	Brown	White	Pink	Black
Scores	24	22	20	18	18	16	15

a Below are the scores for the latest games. Will these scores change the rankings? _____

Team	Red	Blue	Green	Brown	White	Pink	Black
Scores	5	12	7	2	10	6	8

b Complete the new rankings.

Team							
Scores							

2 Roll two 10-sided dice to make a 2-digit number. Write it under 'My number' in the table. Write the numbers that come before and after your number. Repeat this four times.

Number before	My number	Number after

3 Try it with 3-digit numbers. Roll three dice or ask someone to tell you a 3-digit number.

Number before	My number	Number after

Practice

1 Miss Walker asks Leo to solve 6 + 13. Leo starts from number 13. Why? Explain your thinking and show your working out.

2 Use the same strategy to solve these calculations. Show your working out.

a   9 more than 27

b   7 and 19

c   13 + 15

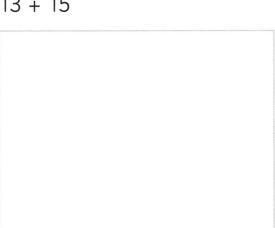

d   8 more than 46

1 Max has 16 toy cars. For his birthday, he was given more as gifts. How many cars might Max have now?

16 + [ ] = [ ]

16 + [ ] = [ ]

16 + [ ] = [ ]

16 + [ ] = [ ]

16 + [ ] = [ ]

2 Roll two 10-sided dice to make a 2-digit number. Record your number. Count on 8 more and write the new number. Repeat this four more times.

[ ] + 8 = [ ]

[ ] + 8 = [ ]

[ ] + 8 = [ ]

[ ] + 8 = [ ]

[ ] + 8 = [ ]

3 Myra's friend gave her some shells. She has 27 shells now. How many might Myra have started with? How many did her friend give her?

[ ] + [ ] = 27

[ ] + [ ] = 27

[ ] + [ ] = 27

OXFORD UNIVERSITY PRESS

1 Julie and Maria play a game. Julie wins by 11 points. What might their scores be?

Julie's score	Maria's score

You might like to use a hundred chart to help you.

---

2 Bill's birthday is in June. Paul's birthday is 14 days after Bill's. When might each of their birthdays be?

Bill's birthday	Paul's birthday

Practice

1 Partition the numbers.

a
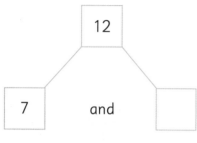

12

7 and ☐

b
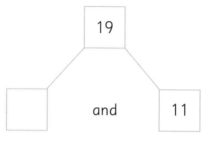

19

☐ and 11

c
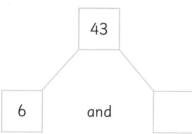

43

6 and ☐

d
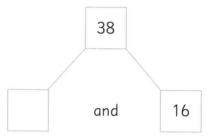

38

☐ and 16

2 The pictures on the left show the same amounts as the pictures on the right. There are 10 sticks in each bundle. Find the matching pairs.

OXFORD UNIVERSITY PRESS

Challenge

1 Find three different ways to partition each number.

Number	Partition 1	Partition 2	Partition 3
13			
18			
21			
27			
35			
62			

2 Amelia had lots of teddy bears. She gave half of them away. How many bears might Amelia have started with? How many did she give away?

Started with	Gave away

1 How many different ways can you partition the number 24?

2 Jack gives nine pens to Ava.
How many pens might Jack
have started with?
How many does he
have now?

You can use this
diagram to help you.

| 9 | and | |

Practice

1 Mr Jones writes 22 − 7 = ☐ on the board.

a Use the number line to solve the problem.

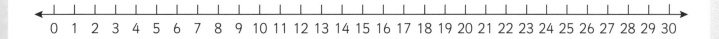

0 1 2 3 4 5 6 7 8 9 10 11 12 13 14 15 16 17 18 19 20 21 22 23 24 25 26 27 28 29 30

b Now solve these problems.

29 − 7 = ☐   34 − 9 = ☐   41 − 11 = ☐

0   10   20   30   40   50   60

2 Bob and Peter play 'First to 30'. Bob wins all five games! Look at how many squares Peter has left uncovered. Work out Peter's score for each of the five games.

> In 'First to 30', each player rolls a dice to make a number. They place that number of counters onto their ten-frames. The first to make 30 by filling three ten-frames wins!

	Squares uncovered	Peter's score (squares covered)
Game 1	3	
Game 2	6	
Game 3	11	
Game 4	14	
Game 5	9	

1 Roll two 10-sided dice. Combine the digits to create a number. Record your number. Count back 7 from your number and write the answer. Repeat this nine more times.

Your number   Answer   Your number   Answer

[ ] – 7 = [ ]   [ ] – 7 = [ ]

[ ] – 7 = [ ]   [ ] – 7 = [ ]

[ ] – 7 = [ ]   [ ] – 7 = [ ]

[ ] – 7 = [ ]   [ ] – 7 = [ ]

[ ] – 7 = [ ]   [ ] – 7 = [ ]

2 Roll a 10-sided dice. Record the number. Count back that many from 26 and write the answer. Repeat this nine more times.

Number rolled   Answer   Number rolled   Answer

26 – [ ] = [ ]   26 – [ ] = [ ]

26 – [ ] = [ ]   26 – [ ] = [ ]

26 – [ ] = [ ]   26 – [ ] = [ ]

26 – [ ] = [ ]   26 – [ ] = [ ]

26 – [ ] = [ ]   26 – [ ] = [ ]

Could you use a number line to help you?

OXFORD UNIVERSITY PRESS

1 Lucas had 32 marbles but he dropped them and lost some. How many marbles might Lucas have lost? How many are left?

2 Alice missed out on 9 points on her mathematics test. Decide what the maximum possible test score could be. Use the counting back strategy to work out Alice's score. Repeat with different maximum scores.

Maximum possible test score	Alice's test score
e.g. 40	31

Practice

You could use blocks to help you, or draw a number line.

1   Find the difference between the two numbers.

a   15 and 7   ☐   b   19 and 13   ☐

c   4 and 12   ☐   d   11 and 23   ☐

e   25 and 9   ☐   f   13 and 29   ☐

2   Choose two numbers. Use the number line to find the difference between your chosen numbers.

My two numbers are _____ and _____

0   100

Challenge

1 Zoe is thinking of two numbers. The difference between her two numbers is 7. What could Zoe's two numbers be?

[ ] and [ ]   [ ] and [ ]

[ ] and [ ]   [ ] and [ ]

[ ] and [ ]   [ ] and [ ]

[ ] and [ ]   [ ] and [ ]

2 Roll two 10-sided dice to make two 2-digit numbers. Find the difference between the first and second numbers. Repeat this five more times.

First number	Second number	Difference
e.g. 56	65	The difference is 9.

You could use a number
line to help you.

+9

50   56   60   65   70

1. The red team played the blue team at football. The red team scored more than double the blue team's score. What might each team have scored? Find the difference between their scores.

Red team's score	Blue team's score	Difference

2. Sofia put 31 candles on her mother's birthday cake. Dad says she is going to be older than that. Decide how old Sofia's mother might be turning and how many more candles Sofia should add.

How old will Sofia's mother be?	How many more candles?
e.g. 35	4 more candles

OXFORD UNIVERSITY PRESS

Practice

1 Grouping items into groups of five can make counting larger collections easier. Count by 5s to find the amounts below.

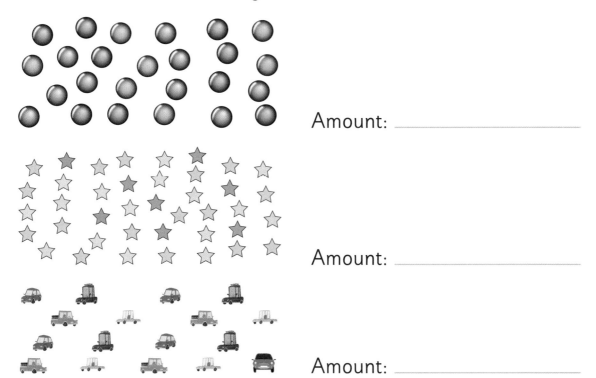

Amount: _____

Amount: _____

Amount: _____

2 Skip counting in 2s can also be a quick way to count. How many eyes are on the people below?

Eyes: _____

Eyes: _____

Eyes: _____

Challenge

1 Skip count to complete the blanks.

a [ ] [ ] [ ] [ ] 40 [ ] 50

b [ ] 15 [ ] [ ] 30 [ ] [ ]

c 18 [ ] [ ] [ ] [ ] 28 [ ]

d [ ] [ ] 45 [ ] [ ] 30 [ ]

e [ ] 70 [ ] [ ] [ ] [ ] 20

- -

2 Make skip counting patterns. Start from a number of your choice.

a Skip count by 2.

b Skip count by 5.

c Skip count by 10.

- -

3 Make two patterns that skip count **backwards**. Decide which numbers to start from.

a Skip count in 2s.

b Skip count in 5s.

1 Mia creates a skip counting pattern that lands on 40. What might Mia's skip counting pattern look like?

Skip count by	Skip counting pattern

2

a A group of people are at the bus stop. How many eyes are there altogether?

Number of people	Skip counting pattern	Total eyes

b A florist sells bunches of flowers with 10 flowers in each bunch. How many flowers are there altogether?

Number of bunches	Skip counting pattern	Total flowers

Practice

1 Draw the problem and work out the answers.

a   10 shared between 2.

b   12 shared between 3.

c   15 shared between 5.

d   16 shared between 4.

2 Isla has 12 sweets to share. How many will each person get if she shares the sweets between:

a   two people?

b   three people?

c   four people?

d   six people?

OXFORD UNIVERSITY PRESS

Challenge

1 Oliver gives his friends four cookies each. How many cookies might Oliver have to share?

How many friends?	Equal shares (draw 4 cookies each)	Total cookies

2

a Can Nico share nine sweets equally between three people? Show your working out.

When you are sharing equally, make sure everyone gets the same amount.

b Can Nico share nine sweets equally between four people? Show your working out.

1 How many people could Isabelle share 20 sweets equally between?

2 Zac gives each friend three sweets. How many sweets might Zac have?

How many friends?	Equal groups (draw)	How many sweets in total?

Practice

1

H
O
M
E

What colour bird is:

a first?

b fourth?

c third?

d sixth?

2

a Draw a black bird between the blue and green birds.

b Circle the position of the black bird: | 1st | 2nd | 3rd | 4th | 5th |

c How many birds are there now?

3 Look at the picture with the black bird drawn in. In what place is the:

a green bird?

b purple bird?

c orange bird?

d blue bird?

Is it always good to be first?

1 Draw:

a the first thing you ate today.

b the second thing you ate today.

c the first thing you did at school today.

d the second thing you did at school today.

2 Look at the building.

a How many floors are there?

b How many of the floors have windows?

c Draw a pot plant on the first floor.

d What is on the third floor?

e What is on the 5th floor?

f How many windows are there on the 4th floor?

1

a   Draw yourself in front of the classroom door.

b   Draw 6 students lined up behind you.

c   Label the people in line from 1st to 7th.

2

a   Draw 3 more carriages on the train.

b   Draw a different number of people in each carriage.

c   Record how many people are in each carriage.

First carriage ☐   Second carriage ☐

Third carriage ☐   Fourth carriage ☐

Practice

1 Circle all the pictures with equal halves.

2 Circle all the pictures with equal quarters.

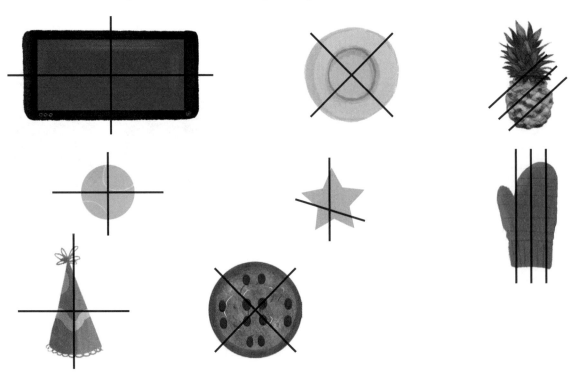

OXFORD UNIVERSITY PRESS

1 Show four different ways the squares could be cut into quarters.

2

a Molly ate $\frac{1}{2}$ of her pizza. What might Molly's pizza look like now?

b Nico ate $\frac{1}{4}$ of his pizza. What might Nico's pizza look like now?

c Frank ate $\frac{3}{4}$ of his pizza. What might Frank's pizza look like now?

If you have eaten $\frac{3}{4}$ of your pizza, how much is left?

1. Design a mat that is $\frac{1}{4}$ red, $\frac{1}{2}$ blue and $\frac{1}{4}$ green.

2. Create your own pizza. What toppings will you use? You can have different toppings on different parts of the pizza.

Fraction   **Topping**

_____ of my pizza has _____

_____ of my pizza has _____

_____ of my pizza has _____

_____ of my pizza has _____

OXFORD UNIVERSITY PRESS

Practice

1 Circle half of each group.

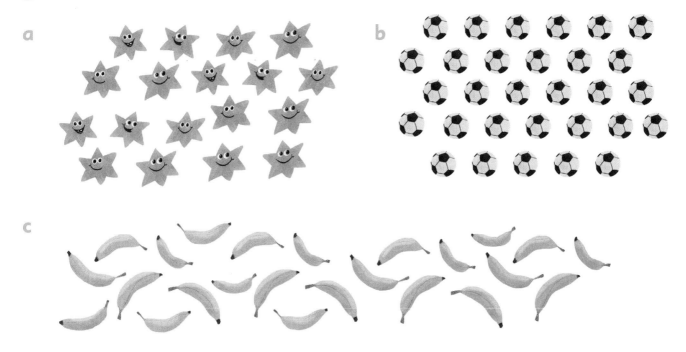

a

b

c

- -

2 Circle one quarter of each group.

a

b

c

How many different ways can you find to show this?

1. Sam and his mother made cookies. Half the cookies were plain and half were chocolate chip. Draw what the cookies might look like.

2. Timmy and Kaden are playing marbles. One quarter of their marbles are blue. Draw what their marbles might look like below. How many different ways can you show this?

1 Work out the answers.

Question	Working out
There are 24 students in Sienna's class. Half are boys. How many boys are there?	
Dan has 20 people in his class. $\frac{1}{4}$ of them are away today. How many are away?	
There are 30 people watching tennis. Half are wearing hats. How many people are wearing hats?	
Jess has 16 people in her class. $\frac{1}{4}$ are away today. How many of her class are at school?	

2 Rose pulled a handful of counters out of a container. $\frac{1}{2}$ were red, $\frac{1}{4}$ were green and the rest were yellow. How many of each colour might there be?

Total counters	Red counters	Green counters	Yellow counters

Practice

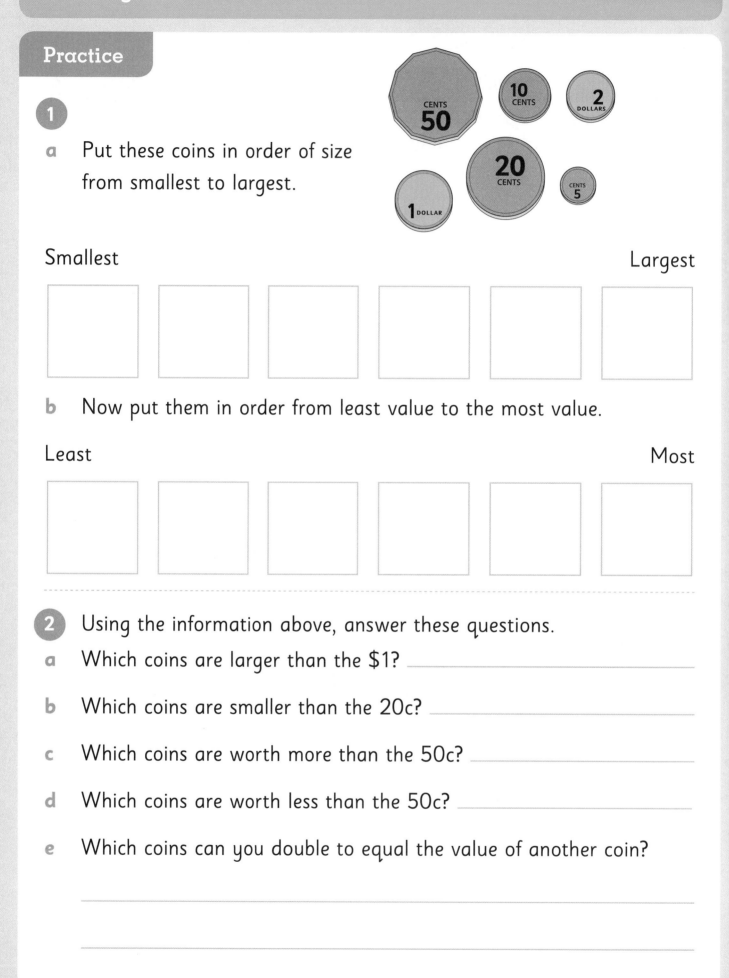

1

a Put these coins in order of size from smallest to largest.

Smallest   Largest

b Now put them in order from least value to the most value.

Least   Most

2 Using the information above, answer these questions.

a Which coins are larger than the $1? _____

b Which coins are smaller than the 20c? _____

c Which coins are worth more than the 50c? _____

d Which coins are worth less than the 50c? _____

e Which coins can you double to equal the value of another coin?

_____

_____

OXFORD UNIVERSITY PRESS

1 What coins from page 38 would you use to buy these items? Draw the coins in order from most to least value.

Item	Draw the coins (most value to least value)
$1.50	
$3.00	
$2.25	
$3.95	
$5.10	

2 Order the coins from most to least value. Add them to find the total.

Coins	Order	Total
20 CENTS, 10 CENTS, 50 CENTS, 5 CENTS		
5 CENTS, 2 DOLLARS, 5 CENTS, 1 DOLLAR, 50 CENTS		
20 CENTS, 1 DOLLAR, 10 CENTS, 20 CENTS, 50 CENTS		
10 CENTS, 1 DOLLAR, 5 CENTS, 20 CENTS		
2 DOLLARS, 1 DOLLAR, 5 CENTS, 50 CENTS, 20 CENTS		

1

a Sometimes you can use more than one coin to make the same value as a single coin. How many coins can you do this with?

b Can you find examples where more than two coins are equal in value to a single coin?

> I can use five 10-cent pieces to make 50 cents!

2 Ben found three coins under his bed. Two were the same and one was different. What might the coins be and how much might Ben have?

Coins	Amount

Practice

1 Continue the patterns.

a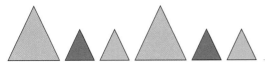

b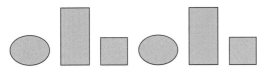

c M W A M W A _____

d P P R P P R _____

---

2 Circle the error in each pattern. Complete the table.

Circle the error in each pattern	What should it be?	Rule for the pattern
⬆⬆⬇⬆⬆⬇⬇⬆		
●●●○●○●○●		
X Y Z X Y Z X X Z		
O U U O O U O O U		
I I J I I I J J I I I J J		

1 Create your own patterns. What is the rule for each pattern?

Rule: _____

Rule: _____

2 These patterns grow each time. Draw the next one in each pattern.

a

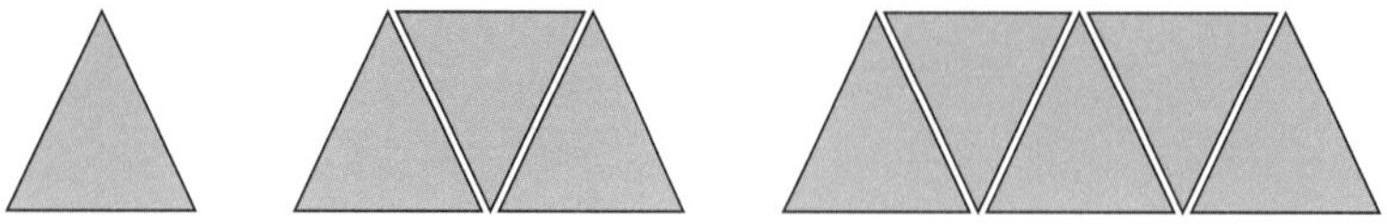

b

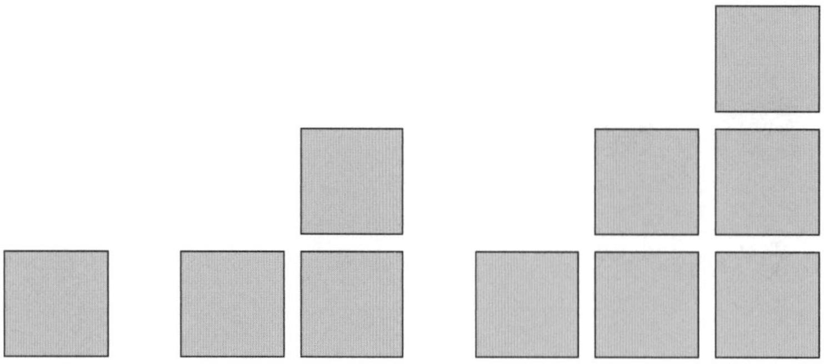

c

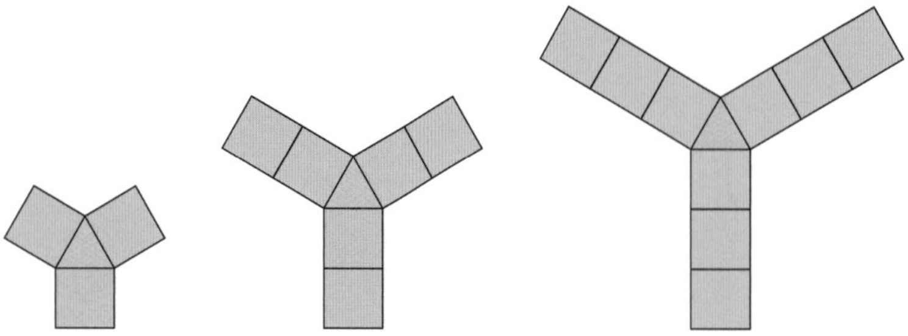

OXFORD UNIVERSITY PRESS

1 The path to Izzy's house has stepping stones that are in a pattern. What might the pattern look like?

> I wonder if it is a shape pattern, a colour pattern or a size pattern?

2 Ben opened a packet of sweets. He grouped the different coloured sweets together. He noticed that they made a growing pattern. What pattern might the sweets make?

Practice

1 Finish the number patterns.

a   36, 38, 40, 42, 44, _____ _____ _____ _____ _____

b   55, 60, 65, 70, 75, _____ _____ _____ _____

c   120, 110, 100, 90, 80, _____ _____ _____ _____ _____

d   80, 78, 76, 74, 72, _____ _____ _____ _____ _____

e   What does each pattern go up or down by?

Pattern a: _____   Pattern b: _____

Pattern c: _____   Pattern d: _____

2 Fill in the missing numbers to complete the patterns.

a   60, 70, ☐ , 90, ☐ , ☐ , 120, 130, ☐

b   110, 105, 100, ☐ , ☐ , 85, ☐ , ☐ , 70

c   92, ☐ , 96, 98, ☐ , ☐ , ☐ , 106, 108

d   190, ☐ , ☐ , 160, ☐ , ☐ , 130, 120

e   What does each pattern go up or down by?

Pattern a: _____   Pattern b: _____

Pattern c: _____   Pattern d: _____

Challenge

1

a Colour in the hundred chart to show a pattern.

b What is your pattern?

_____

c What number comes next in your pattern?

_____

1	2	3	4	5	6	7	8	9	10
11	12	13	14	15	16	17	18	19	20
21	22	23	24	25	26	27	28	29	30
31	32	33	34	35	36	37	38	39	40
41	42	43	44	45	46	47	48	49	50
51	52	53	54	55	56	57	58	59	60
61	62	63	64	65	66	67	68	69	70
71	72	73	74	75	76	77	78	79	80
81	82	83	84	85	86	87	88	89	90
91	92	93	94	95	96	97	98	99	100

2

a Colour a pattern that counts in 10s but does **not** start at 10.

1	2	3	4	5	6	7	8	9	10
11	12	13	14	15	16	17	18	19	20
21	22	23	24	25	26	27	28	29	30
31	32	33	34	35	36	37	38	39	40
41	42	43	44	45	46	47	48	49	50
51	52	53	54	55	56	57	58	59	60
61	62	63	64	65	66	67	68	69	70
71	72	73	74	75	76	77	78	79	80
81	82	83	84	85	86	87	88	89	90
91	92	93	94	95	96	97	98	99	100

b Colour a pattern that counts in 5s but does **not** start at 5.

1	2	3	4	5	6	7	8	9	10
11	12	13	14	15	16	17	18	19	20
21	22	23	24	25	26	27	28	29	30
31	32	33	34	35	36	37	38	39	40
41	42	43	44	45	46	47	48	49	50
51	52	53	54	55	56	57	58	59	60
61	62	63	64	65	66	67	68	69	70
71	72	73	74	75	76	77	78	79	80
81	82	83	84	85	86	87	88	89	90
91	92	93	94	95	96	97	98	99	100

1

a   One of the numbers in Mo's pattern is 15. What might Mo's pattern look like?

b   One of the numbers in Kim's pattern is 28. What might Kim's pattern look like?

2   House numbers in a street often go up in 2s. The numbers on Jack's street make a different pattern. What might the house numbers on Jack's street look like?

My house is number 18. Do you know what number your house is?

Practice

1 Here are five different-sized items.

Order the items from which is shortest to longest in real life.

2 Find the area of the following shapes.

a _____ tiles

b _____ tiles

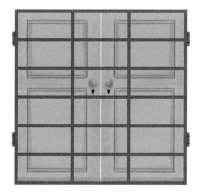

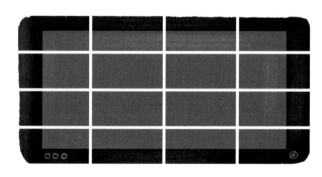

c _____ tiles

d _____ tiles

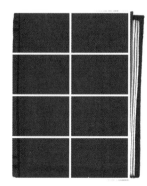

1. Use paperclips to estimate and measure the length of these items.

Item	Estimate	Measure

2. Use blocks to estimate and measure the area of these items.

Item		Estimate	Measure
A box lid			
A poster			
A wall			

How close were your estimates? _____

OXFORD UNIVERSITY PRESS

1 Find items that are shorter than, the same length as, or longer than this book.

Shorter than	Same length as	Longer than

2 How many different shapes can you create with an area of nine squares?

Practice

1 Find the volume of these objects.

a _____ cubes

b _____ cubes

c _____ cubes

d _____ cubes

e Order them from smallest to largest.

2 Look at these containers.

Draw them in order from the container that has the most capacity to the least capacity.

OXFORD UNIVERSITY PRESS

Challenge

1 Match the picture with its correct volume in cubes.

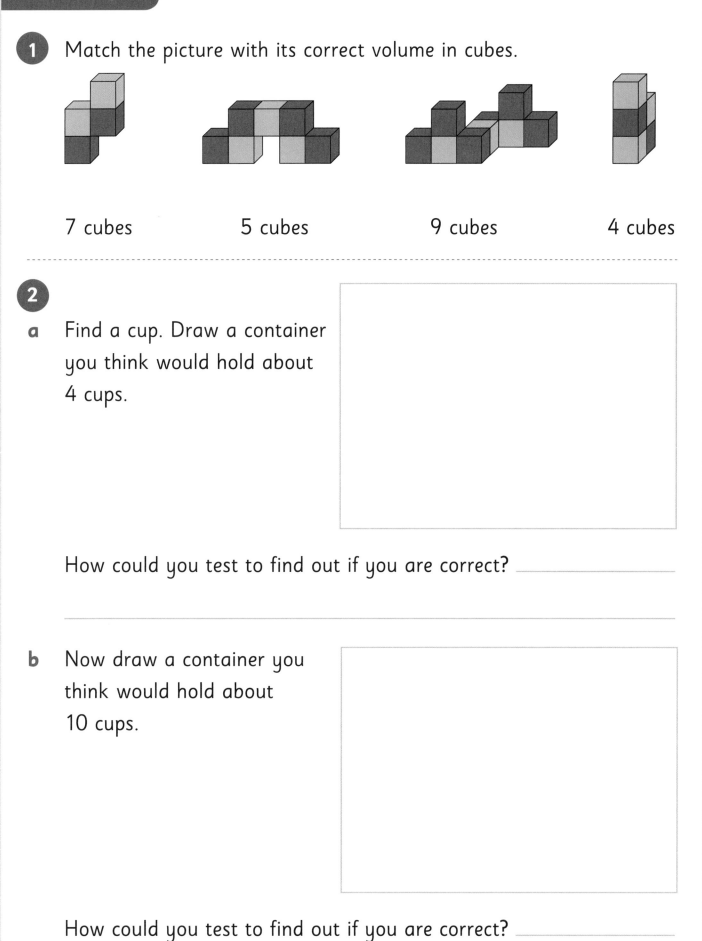

7 cubes   5 cubes   9 cubes   4 cubes

2

a Find a cup. Draw a container you think would hold about 4 cups.

How could you test to find out if you are correct? _____

_____

b Now draw a container you think would hold about 10 cups.

How could you test to find out if you are correct? _____

_____

1 Draw different objects, all with a volume of 8 cubes.

2

a Draw some containers with a larger capacity than a coffee cup.

b Draw some containers with a smaller capacity than a coffee cup.

How many cups might it take to fill a bathtub?

OXFORD UNIVERSITY PRESS

Practice

1 Circle the heavier item in each pair.

a   or

b    or

c    or

d    or

2 Find five items that are heavier and five items that are lighter than this mathematics book.

Heavier	Lighter

1

a Draw five different items you can see around you.

b Order your five items from lightest to heaviest.

2 Draw pairs of items that you think weigh about the same.

a

b

c

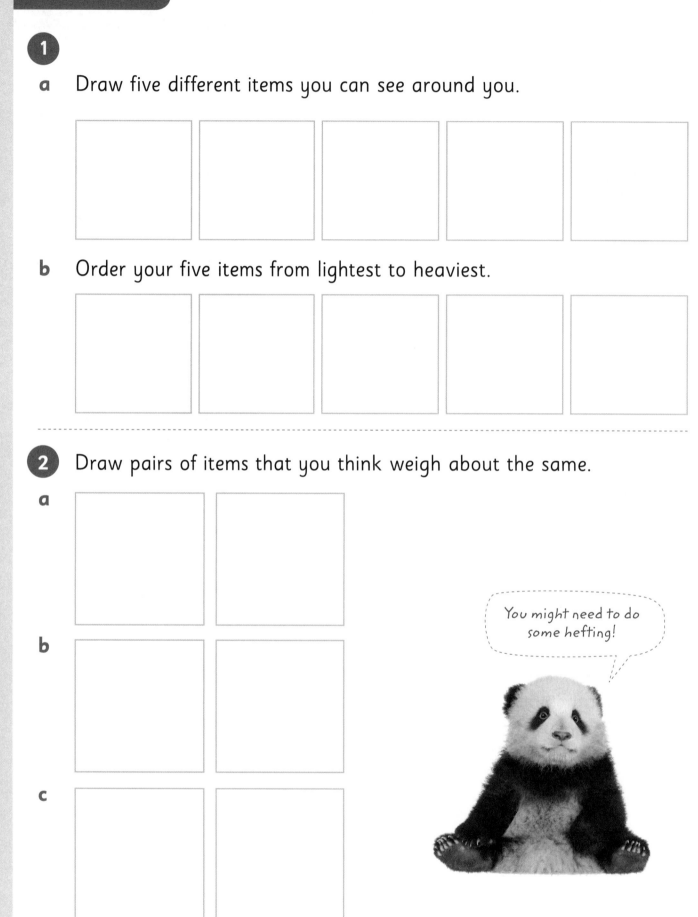

You might need to do some hefting!

1 Draw an item on each side of the balance scales to make these true.

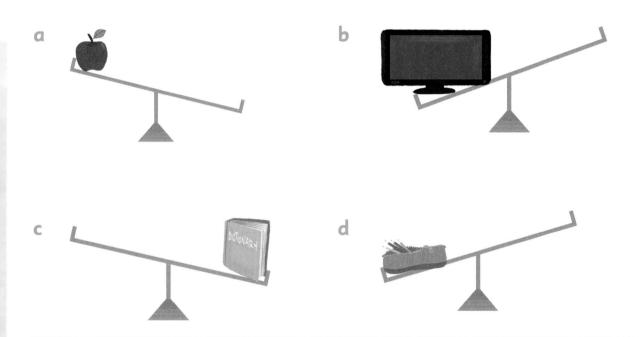

a

b

c

d

- -

2 One glue stick weighs the same as 4 cubes and one eraser weighs 3 cubes. Use this information and the pictures to answer the questions.

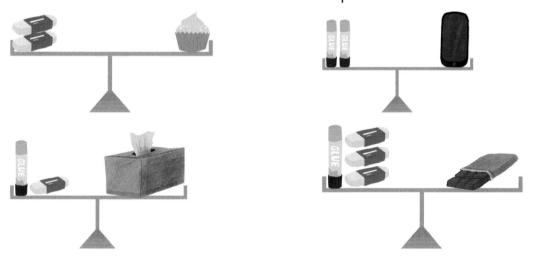

a How many cubes does the cake weigh? _____ cubes

b How many cubes does the phone weigh? _____ cubes

c How many cubes do the tissues weigh? _____ cubes

d How many cubes does the chocolate weigh? _____ cubes

Practice

1 Match the analogue clock time to the digital clock time.

2 Write the time in numbers, words and draw the clock hands.

Digital time	The time in words	Analogue time
11:30		
:		
:	nine o'clock	
:		
:	half past one	

OXFORD UNIVERSITY PRESS

1   What times are these clocks showing?

a  _____

b  _____

c  _____

d  _____

e  _____

f  _____

2   Draw the times on the clocks.

a   10.30

b   5 o'clock

c   3.30

d   12.30

e   6 o'clock

f   7.30

a Show your favourite time of the day on the two clocks.

b What would you be doing at this time of the day?

_____

_____

2 One of the hands has fallen off the clock. The hand that is left is pointing at the 6. What might the time be?

 OXFORD UNIVERSITY PRESS

Practice

1 Put these events in order from shortest to longest duration.

a   Have a drink

b   Watch a movie

c   A day at school

d   Eat dinner

2 Would you measure the time until these events in hours, days, weeks, months or years?

Event		Time measured in
Your next birthday		
You finish school		
The weekend		
Winter		

1

a Draw an event that is coming up soon.

b What is the event? When will it happen?

_____

_____

- -

2 Match the event with the time you would measure it in.

Weekend at the beach  Hours

The school holidays  Minutes

 A school day Weeks

Play on a seesaw  Days

OXFORD UNIVERSITY PRESS

1 Order the birthdays of the people in your family.

Don't forget to include your own birthday!

2 Draw some things you could do in:

a one hour.

b one minute.

Practice

1 Here are some 2D shapes.

a Sort the shapes into groups.

b How did you sort the shapes? Give each group a heading.

_____

c How many different groups did you have? _____

2

a Draw two parallel lines.

b Draw two lines that are **not** parallel.

c What does parallel mean?

_____

OXFORD UNIVERSITY PRESS

1 Guess my shape.

a I have _____ corners. I have two short and two long sides.

What am I? _____

b Write your own 'guess my shape' question for someone to solve.

```



```

What am I? _____

2 Draw some 2D shapes and describe their features.

Shape	Description of features

1   Draw a picture using only shapes. Use as many different shapes as you can!

2   Label the shapes in your picture. Record below how many of each shape you used.

Shape	How many in the picture?

OXFORD UNIVERSITY PRESS

Practice

1 Draw lines to match each 3D shape to its name.

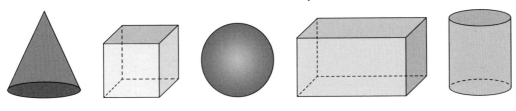

| cube | rectangular prism | cone | sphere | cylinder |

2 Draw and name the 3D shapes that match these descriptions.

Description	Draw and name the 3D shape
6 faces	
0 edges	
5 faces	
6 corners	
9 edges	

1 Draw real-life objects that are the same shape as these 3D shapes.

3D shape	Real-life object
sphere	
cone	
cylinder	
cube	
rectangular prism	

2

a Sort these 3D shapes into groups. Give each group a heading. Then, under the correct heading, draw each 3D shape **or** write its name.

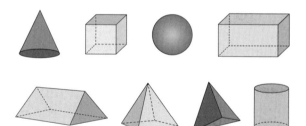

b How did you sort the 3D shapes? _____

_____

OXFORD UNIVERSITY PRESS

1 Guess which 3D shape I am.

a I have _____ corners. I have five faces.

What am I? _____

b Write a 'guess which 3D shape I am' question for someone to solve.

What am I? _____

2 Draw some 3D shapes. Write which 2D shapes make up each 3D shape.

Did you know that a cube is made up of six squares?

3D shape	2D shapes that make up the 3D shape

Practice

1 Look at the picture and answer the questions.

a What is on the rug? _____

b What is under the bed? _____

c The photo frame is _____ the shelves.

d The rug is _____ the bed.

e The train is _____ the shelves.

2 Follow these directions to create a picture.

a Draw a sun in the top right-hand corner.

b Draw two flowers in the bottom middle.

c Draw a person next to the flowers.

d Draw a pond in the bottom left corner.

e Draw three ducks on the pond.

f Draw a butterfly above the person.

Challenge

1 Use the picture to describe the position of the items.

Item	Position description
shelves	
teddy bear	
clock	
window	

2 Add some items to the picture.

a Draw a table above the rug.　**b** Add a vase next to the train.

c Put a pillow on top of the bed.　**d** Draw a lamp on the table.

1 Look at the classroom map and answer the questions.

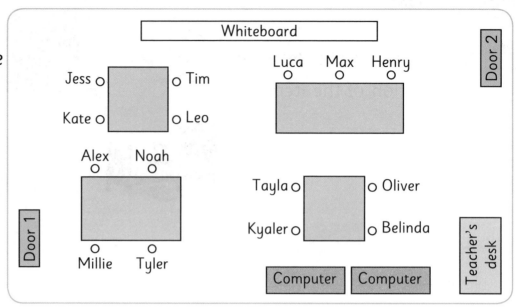

Whiteboard

Jess ○ ☐ ○ Tim
Kate ○ ○ Leo

Luca   Max   Henry
○   ○   ○

Alex   Noah
○   ○

Tayla ○ ☐ ○ Oliver
Kyaler ○ ○ Belinda

○   ○
Millie   Tyler

Door 1

Door 2

Teacher's desk

Computer   Computer

a Who sits furthest from door 1? _____

b Who sits closest to the teacher's desk? _____

c Who sits next to Leo? _____

d Write a question of your own for someone else to answer.

_____

2 Look at the picture below.

Write the step-by-step instructions you would tell someone, so they could draw the same picture.

What language will you use? You might like to use 'next to', 'under' or 'above'.

_____
_____
_____

Practice

1 Circle the objects below that have been turned anti-clockwise.

2 Look at the map. Work out how many grid spaces from one place to another. For example, to get from school to the shop, go 2 spaces up and 2 spaces right.

How many spaces from:

a   the tree to the pond? _____

b   the house to the swimming pool? _____

c   the playground to the sports field? _____

d   the forest to the shop? _____

Challenge

1 Use the map and follow the directions. Where do you finish?

Directions:

a Start at the tree.

b Go right 3 spaces.

c Go down 2 spaces.

d Go right 2 spaces.

e Go down 2 spaces.

f Go left 3 spaces.

g Where did

you finish? _____

2

a Using the map above, write your own set of directions for someone else to follow.

_____

_____

_____

b Where did they finish? Were your directions correct?

_____

_____

_____

1   Draw a map on the grid. You should be able to use your map to follow these directions.

a   Start at Leo's house in the bottom left corner and go 4 spaces right.

b   Go up 3 spaces to Zoe's house.

c   Go left 1 space and then up 2 spaces to Noah's house.

d   Go right 4 spaces to Jack's house.

e   Go down 3 spaces and left 1 space to Claire's house.

2

a   Make a region on your map with shops in it.

b   Draw a pathway from one of the shops to Claire's house.

c   Add in a sports field with a boundary around it.

Practice

1 On the graph, show how many people have each hair colour.

Number of people

	15				
14				
13				
12				
11				
10				
9				
8				
7				
6				
5				
4				
3				
2				
1				

Blonde   Red   Black   Brown

Hair colour

2 The graph shows how many people like each sport. Use the graph to complete the table.

Sport	How many?

9
8
7
6
5
4
3
2
1

Tennis Basketball Swimming Netball Football Baseball

OXFORD UNIVERSITY PRESS

Challenge

1 Class 1 was asked what sports they play outside school. These are their answers.

Put the data into the table below.

You may want to use tally marks!

Aidan	Basketball	Lucy	Netball
Josh	Football	Daniel	Basketball
Ava	Dancing	Holly	Dancing
Mo	Football	Sam	Karate
Charlie	Basketball	Chase	Football
Liam	Basketball	Ash	Dancing
Mya	Netball	Luca	Dancing
Laura	Football	Lily	Football
Logan	Football	Jake	Karate
Ella	Dancing	Jayda	Netball
Riley	Basketball	Harper	Football
Nico	Football	Jess	Netball

Sport	Number of people

2 Use the table to draw a pictograph.

1 Now it is your turn to collect some data. What are some good questions you could ask people?

2 Choose one of your questions and collect the data.

a Your question _____

b Record people's answers. Tally how many people give each answer.

Answers	Tally marks

Great work! Now you can create a graph from the data you have collected.

OXFORD UNIVERSITY PRESS

Practice

1 Answer the questions about the graph.

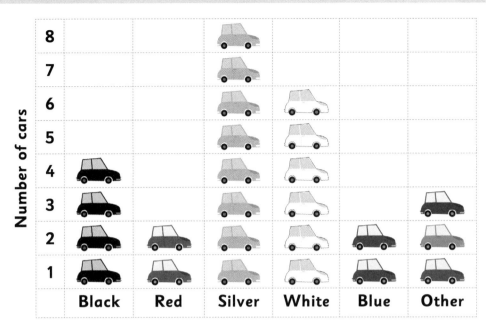

a Which is the most popular colour car?

_____

b Which two colours have the same number of cars? _____

c How many cars were there in total? _____

d What might go in the 'other' column? _____

2 Use the data to fill in the graph.

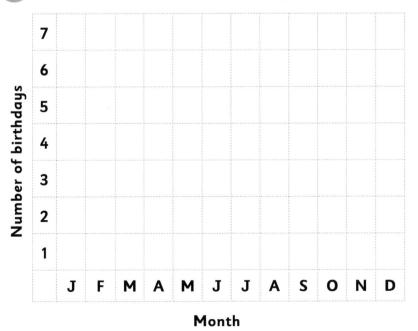

Month	Number of birthdays
January	II
February	III
March	I
April	
May	IIII
June	I
July	III
August	IIII
September	IIII
October	
November	III
December	I

Number of people in our families

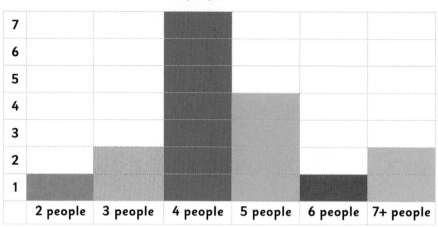

1 Look at the graph. Write four questions you could ask someone about this graph.

1 _____

2 _____

3 _____

4 _____

2 Here is a completed graph.

a What do you think this graph is about?

_____

_____

b Write two statements about what the graph is showing you.

_____

_____

_____

c What title would you give this graph? _____

_____

OXFORD UNIVERSITY PRESS

1 Use this information to complete the graph.

- The most popular colour is blue with 7 people.

- Two more people liked blue than red.

- Two people liked yellow.

- The least popular colour was green.

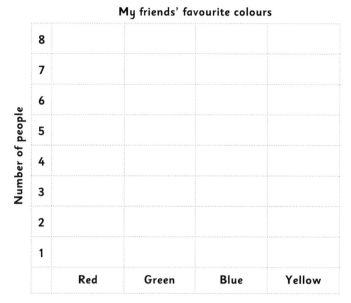

My friends' favourite colours

2 Below are Rahul's answers about the graph. Can you work out what the questions might have been?

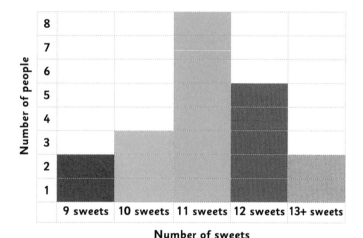

Number of sweets

Answer	Question
8 people	
9 sweets	
9 and 13+ sweets	
7 people	
20 people	

Practice

1 Write **certain**, **impossible** or **maybe** to show the chance of each event happening.

A cat will read the newspaper.

It will get dark.

It will be a sunny day.

You will make a new friend.

_____

_____

_____

_____

2

a Draw three things that are **certain** to happen today.

b Draw three things that are **impossible** to happen today.

Challenge

1 For each bag, what is the chance of pulling out a blue marble? Write **certain**, **impossible** or **maybe**.

Bag of marbles	Chance

2 Draw groups of marbles to match the descriptions.

Description	Marbles
It is impossible to pick a yellow marble.	
It is certain you will pick a blue marble.	
You will maybe pick a green marble.	
Red has the most chance of being picked.	

1. Tatsuo is 7 years old. He lives near a beach in Japan and spends his weekdays at school.

   a Write two things that are impossible to happen to Tatsuo today.

   _____

   _____

   _____

   b Write two things that could maybe happen to Tatsuo today.

   _____

   _____

   _____

   c Write two things that are certain to happen to Tatsuo today.

   _____

   _____

   _____

2.

   a Flip a coin six times. Predict what the coin will land on (heads or tails) before you flip it each time. Record your answers below.

Prediction	Actual result

   b How many times were you right? _____

   c How many times were you wrong? _____

   d Describe the chance of getting it right. _____

OXFORD UNIVERSITY PRESS

ANSWERS

UNIT 1: Topic 1

Practice

1 a | 41 | **42** | 43 | **44** | 45 | 46 | **47** | **48** | **49** | 50 | 51

b | 86 | 87 | **88** | **89** | 90 | **91** | 92 | **93** | 94 | 95 | **96**

c | 63 | **62** | **61** | **60** | 59 | 58 | **57** | **56** | **55** | 54 | 53

d | **99** | 98 | 97 | **96** | 95 | 94 | **93** | **92** | 91 | **90** | 89

2

1					6				
								19	
		33							
			45						
					67				
71									
	82							90	
								100	

Challenge

1 a 24   b 19   c 12
   d 21   e 7

2 Teacher to check. The difference should be 14. For example, 16 and 2, 24 and 10, 45 and 31.

Mastery

1 Teacher to check. Must be 6 numbers that add to 30, e.g. 4 + 5 + 8 + 4 + 3 + 6

2 a   20, 24, 42, 46, 64, 68, 86
   b   25, 47, 63, 69, 85

UNIT 1: Topic 2

Practice

1 twenty-three, seventeen, forty-one, sixty-six

2

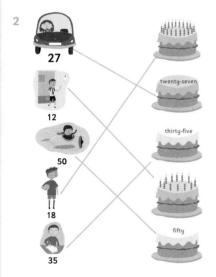

Challenge

1 Teacher to check.

2 Teacher to check.

3

Number line	Number (numerals)	Number (words)
18 19 20 ⨯ 22 23 24	21	twenty-one
34 35 36 37 ⨯ 39 40	38	thirty-eight
27 ⨯ 25 24 23 22 21	26	twenty-six
43 42 ⨯ 40 39 38 37	41	forty-one
89 ⨯ 91 92 93 94 95	90	ninety

Mastery

1 Teacher to check.

2 16 sixteen

   25 twenty-five

   34 thirty-four

   43 forty-three

   52 fifty-two

   61 sixty-one

   70 seventy

UNIT 1: Topic 3

Practice

1

```
              39 43      54
 |--|--|--|--|--|--|
 0  10 20 30 40 50 60

   51        66      79
 |--|--|--|--|--|
 50  60  70  80  90
```

2 Teacher to check.

Challenge

1 a–f Teacher to check.

2 a 27, 57, 24, 54, 52, 25, 42, 72, 45, 74, 47, 75

   b Largest to smallest: 24, 25, 27, 42, 45 ,47, 52, 54, 57, 72, 74, 75

Mastery

1 a Yes, the scores will change the rankings.

b

Team	Scores
Blue	34
Red	29
White	28
Green	27
Black	23
Pink	22
Brown	20

2 Teacher to check.

3 Teacher to check.

UNIT 1: Topic 4

Practice

1 Teacher to check.

2 a 36   b 26   c 28   d 54

Challenge

1 Teacher to check.

2 Teacher to check.

3 Teacher to check.

Mastery

1 Teacher to check. Julie must have 11 points more than Maria, e.g. Julie 24 and Maria 13.

2 Teacher to check. Paul's birthday must be 14 days after Bill's, e.g. Bill on 2 June and Paul on 16 June.

UNIT 1: Topic 5

Practice

1 a

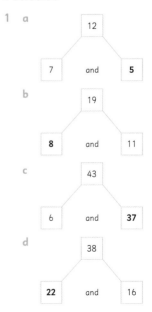

ANSWERS

2

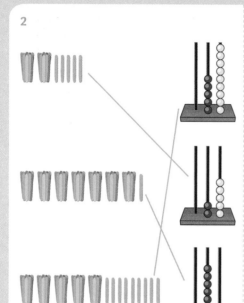

Challenge

1 Teacher to check. The partitioned numbers must add up to the number in the 'Number' column, e.g. 13 = 10 and 3, 5 and 8, or 6 and 7.

2 Teacher to check. Answers must be half the number she started with.

Mastery

1 Teacher to check. Examples include 22 and 2, 10 and 14, 12 and 12, 18 and 6, 13 and 11, 9 and 15.

2 Teacher to check. The difference must be 9, e.g. 24 and 15, 30 and 21, 10 and 1.

UNIT 1: Topic 6

Practice

1 a Teacher to check that 15 has been positioned correctly on the number line.

  b 29 – 7 = 22, 34 – 9 = 25, 41 – 11 = 30

2

Squares uncovered	Peter's score (squares covered)
3	27
6	24
11	19
14	16
9	21

Challenge

1 Teacher to check.

2 Teacher to check.

Mastery

1 Teacher to check. For example, 32 – 8 = 24

2 Teacher to check. For example, 50 – 9 = 41, 45 – 9 = 36, 30 – 9 = 21

UNIT 1: Topic 7

Practice

1 a 8   b 6   c 8
   d 12   e 16   f 16

2 Teacher to check. Students choose two numbers and use the number line to add or subtract to accurately find the difference.

Challenge

1 Teacher to check. The difference needs to be 7, e.g. 23 and 16.

2 Teacher to check.

Mastery

1 Teacher to check. The red team's score needs to be more than double the blue team's score.

2 Teacher to check. For example, 35 is 4 more candles.

UNIT 1: Topic 8

Practice

1 Amount: 25, Amount: 40, Amount: 17

2 Eyes: 24, Eyes: 46, Eyes: 38

Challenge

1 a **20, 25, 30, 35**, 40, **45**, 50
   b **10**, 15, **20, 25**, 30, **40, 45**
   c 18, **20, 22, 24, 26**, 28, **30**
   d **55, 50**, 45, **40, 35**, 30, **25**
   e **80**, 70, **60, 50, 40, 30**, 20

---

2 a Teacher to check. Skip count by 2.
   b Teacher to check. Skip count by 5.
   c Teacher to check. Skip count by 10.

3 a Teacher to check. Skip count backwards by 2.
   b Teacher to check. Skip count backwards by 5.

Mastery

1 Teacher to check. For example, skip count by 5s: 30, 35, 40, 45, 50.

2 a Teacher to check. Skip count in 2s.
   b Teacher to check. Skip count in 10s.

UNIT 1: Topic 9

Practice

1 a 5   b 4   c 3   d 4
2 a 6   b 4   c 3   d 2

Challenge

1 Teacher to check. Equal shares of 4 each, e.g. 3 friends = 4 + 4 + 4 = 12.

2 a Yes, they get 3 each. Teacher to check working out.
   b No. Teacher to check working out.

Mastery

1 Teacher to check: 20, 10, 5, 4, 2, 1.

2 Teacher to check. They must get 3 each, e.g. 5 friends = 3 + 3 + 3 + 3 + 3 = 15.

UNIT 1: Topic 10

Practice

1 a green   b purple
   c red   d grey

2 a Teacher to check. Students should have drawn a black bird in second place (between the green and the blue bird).
   b 2nd
   c 7

3 a first
   b fifth
   c sixth
   d third

OXFORD UNIVERSITY PRESS

Challenge

1 a–d Teacher to check. Look for students who identify appropriate items that are likely to fit the ordinal criteria.

2 a 6

   b 3

   c Teacher to check.

   d A cat

   e A dog

   f 2

Mastery

1 a–c Teacher to check. Look for students who understand how ordinal numbers work and who show fluency with applying ordinal numbers in a practical situation.

2 a Teacher to check. Look for students who demonstrate understanding of cardinal numbers through accurate counting.

   b–c Teacher to check. Answers to c will depend on how many people students have drawn in each carriage.

UNIT 2: Topic 1

Practice

1

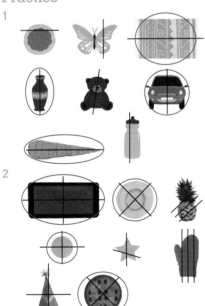

2

Challenge

1 Possible answers:

2 a Possible answers:

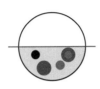

   b Possible answers:

   c Possible answers:

Mastery

1 Teacher to check.

2 Teacher to check.

UNIT 2: Topic 2

Practice

1 a Half of each group should be circled, e.g.

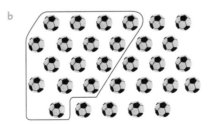

   b

   c

2 a One quarter of each group should be circled, e.g.

   b

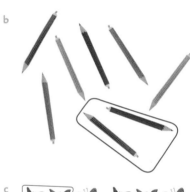

   c

Challenge

1 Teacher to check. Must be half of each type of cookie.

2 Teacher to check. One quarter of the marbles must be blue.

Mastery

1

Question	Working out
There are 24 students in Sienna's class. Half are boys. How many boys are there?	12 (Teacher to check working out.)
Dan has 20 people in his class. 1/4 of them are away today. How many are away?	5 (Teacher to check working out.)
There are 30 people watching tennis. Half are wearing hats. How many people are wearing hats?	15 (Teacher to check working out.)
Jess has 16 people in her class. 1/4 are away today. How many of her class are at school?	12 (Teacher to check working out.)

2 Teacher to check. For example:

Total counters	Red counters	Green counters	Yellow counters
20	10	5	5
16	8	4	4

ANSWERS

UNIT 3: Topic 1

Practice

1 a

b (coins images)

2 a 20c and 50c

  b $1, 10c, $2, 5c

  c $1, $2

  d 20c, 10c, 5c

  e 10c × 2 = 20c,   5c × 2 = 10c,
    $1 × 2 = $2,   50c × 2 = $1

Challenge

1 One possible solution:

Item	Draw the coins (most value to least value)
$1.50	(coins)
$3.00	(coins)
$2.25	(coins)
$3.95	(coins)
$5.10	(coins)

2

Coins	Order	Total
(coins)	(coins)	85c
(coins)	(coins)	$3.60
(coins)	(coins)	$2.00
(coins)	(coins)	$1.35
(coins)	(coins)	$3.75

Mastery

1 a 10c × 2 = 20c, 5c × 2 = 10c,
     $1 × 2 = $2, 50c × 2 = $1

  b Teacher to check. For example,
     10c × 5 = 50c, 10c × 10 = $1,
     50c × 4 = $2

2 Teacher to check. Two of the three
  coins must be the same.

UNIT 4: Topic 1

Practice

1 a (triangles pattern)

  b (circle/square pattern)

  c M W A M W A M W A M W

  d P P R P P R P P R P P

2

Circle the error in each pattern	What should it be?	Rule for the pattern
(arrows)	⬆	2 arrows up and 1 down
(circles)	⬤	Red circle, blue circle
X Y Z X Y Z X (X) Z	Y	X, Y, Z
O (U) U O O U O O U	O	O, O, U
I I J (I) I I J J I I J J	J	I, I, J, J

Challenge

1 Teacher to check.

2 a (triangle shapes)

  b (square shapes)

  c (shapes)

Mastery

1 Teacher to check.

2 Teacher to check.

UNIT 4: Topic 2

Practice

1 a 36, 38, 40, 42, 44, **46, 48, 50,
     52, 54**

  b 55, 60, 65, 70, 75, **80, 85, 90,
     95, 100**

  c 120, 110, 100, 90, 80, **70, 60,
     50, 40, 30**

  d 80, 78, 76, 74, 72, **70, 68, 66,
     64, 62**

  e Pattern a: goes up in 2s, Pattern b:
     goes up in 5s, Pattern c: goes down
     in 10s, Pattern d: goes down in 2s.

2 a 60, 70, **80**, 90, **100, 110**, 120,
     130, **140**

  b 110, 105, 100, **95, 90**, 85, **80,
     75**, 70

  c 92, **94**, 96, 98, **100, 102, 104**,
     106, 108

  d 190, **180, 170**, 160, **150, 140**,
     130, 120

  e Pattern a: goes up in 10s, Pattern b:
     goes down in 5s, Pattern c: goes up in
     2s, Pattern d: goes down in 10s.

Challenge

1 a Teacher to check.

2 a Teacher to check. Pattern must be
     counting by 10s, not starting from 10.

  b Teacher to check. Pattern must be
     counting by 5s, not starting from 5.

Mastery

1 a–b Teacher to check.

2 Teacher to check.

UNIT 5: Topic 1

Practice

1 The correct order is:

  sharpener

  pencil

  banana

  hammer

  table

OXFORD UNIVERSITY PRESS

2 a 18 tiles   **b** 16 tiles
  c 8 tiles   **d** 12 tiles

Challenge

1 Teacher to check.

2 Teacher to check.

Mastery

1 Teacher to check.

2 Teacher to check. For example:

UNIT 5: Topic 2

Practice

1 a 5 cubes   **b** 8 cubes
  c 9 cubes   **d** 20 cubes
  e Smallest to largest: a, b, c, d
    (or 5, 8, 9, 20).

2

Challenge

1

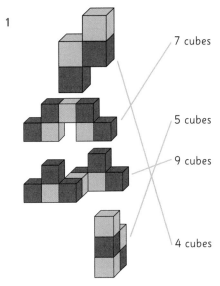

7 cubes

5 cubes

9 cubes

4 cubes

2 a Teacher to check.
  b Teacher to check.

Mastery

1 Teacher to check. Objects should
  each have a volume of 8 cubes, e.g.

2 a–b Teacher to check.

UNIT 5: Topic 3

Practice

1 a   **b**

  c   **d**

2 Teacher to check.

Challenge

1 a Teacher to check.
  b Do children accurately arrange their
    items from lightest to heaviest?

2 a–c Teacher to check.

Mastery

1 a–d Teacher to check.

2 a 6 cubes   **b** 8 cubes
  c 7 cubes   **d** 13 cubes

UNIT 5: Topic 4

Practice

1

 to

 to

  to

 to

2

Digital time	The time in words	Analogue time
11:30	eleven-thirty	
6:00	six o'clock	
9:00	nine o'clock	
5:30	five-thirty	
1:30	half past one	

Challenge

1 a 2.30
  b 11.00
  c 9.30
  d 4.30
  e 2.00
  f 8.30

2 a   **b**   **c**

  d   **e**   **f**

Mastery

1 a–b Teacher to check.

2 Teacher to check. For example,
  6.00 or anything half-past the hour.

ANSWERS

UNIT 5: Topic 5

Practice

1 a, d, c, b

2

Event	Time measured in
Your next birthday	Teacher to check. Days, weeks or months.
You finish school	Teacher to check. Likely answer is years.
The weekend	Teacher to check. Hours or days.
Winter	Teacher to check. Days, weeks or months.

Challenge

1 **a–b** Teacher to check.

2 Weekend at the beach = days

  The school holidays = weeks

  A school day = hours

  Play on a seesaw = minutes

Mastery

1 Teacher to check.

2 **a–b** Teacher to check.

UNIT 6: Topic 1

Practice

1 Teacher to check. For example, group shapes with more or fewer than 3 sides.

2 **a–b** Teacher to check.

  c 2 lines, side by side, with the same continuous distance between them.

Challenge

1 **a** 4 corners. I am a rectangle.

  b Teacher to check.

2 Teacher to check. Look for descriptions such as the number of corners or sides, and whether sides are parallel.

Mastery

1 Teacher to check.

2 Teacher to check.

UNIT 6: Topic 2

Practice

1

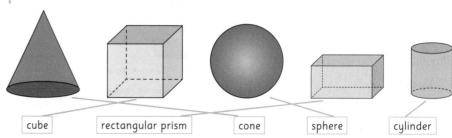

| cube | rectangular prism | cone | sphere | cylinder |

2 Teacher to check. Examples of possible answers are below. Check drawings.

Description	Draw and name the 3D shape
6 faces	cube
0 edges	sphere
5 faces	pyramid
6 corners	triangular prism
9 edges	triangular prism

Challenge

1 Teacher to check. Examples of possible answers are:

3D shape	Real-life object
sphere	ball, marble
cone	party hat, ice-cream cone
cylinder	drink can, candle
cube	dice, ice cube
rectangular prism	tissue box, brick

2 **a–b** Teacher to check. For example,

5 or more faces	Fewer than 5 faces
cube	cone
rectangular prism	sphere
triangular prism	cylinder
square-based pyramid	
triangular-based pyramid	

Mastery

1 **a** I have 5 corners. I am a square-based pyramid.

  b Teacher to check.

2 Teacher to check, for example:

3D shape	2D shapes that make up the 3D shape
cube	6 squares
square-based pyramid	1 square, 4 triangles

UNIT 7: Topic 1

Practice

1 **a** book   **b** shoes

  c above   **d** next to

  e on top of

2 **a–f** Teacher to check.

Challenge

1 Answers may vary but could include:

Item	Position description
shelves	next to the rug
teddy bear	on top of the bed
clock	above the bed
window	next to the clock

2 **a–d** Teacher to check.

Mastery

1 **a** Henry   **b** Belinda

  c Tim   **d** Teacher to check.

2 Teacher to check.

UNIT 7: Topic 2

Practice

1

2 **a** 4 spaces right

  b 3 spaces down and 1 right (or 1 right, 3 down)

  c 1 space down and 3 right (or 3 right, 1 down)

  d 2 left and 2 up (or 2 up, 2 left)

OXFORD UNIVERSITY PRESS

Challenge

1   At the playground.

2   a–b   Teacher to check.

Mastery

1   a–e   Teacher to check.

2   a–c   Teacher to check.

UNIT 8: Topic 1

Practice

1
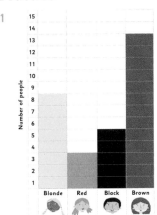

Hair colour

2

Sport	How many?
Tennis	III
Basketball	IIII II
Swimming	II
Netball	IIII
Football	IIII IIII
Baseball	I

Challenge

1

Sport	No. of people
Basketball	5
Netball	4
Karate	2
Football	8
Dancing	5

2   Teacher to check. Graph may look like this:

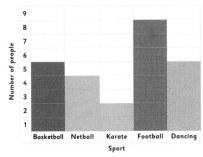

Mastery

1   Teacher to check.

2   a–b   Teacher to check.

UNIT 8: Topic 2

Practice

1   a   silver

   b   red and blue

   c   25

   d   Car colours that don't have their own column on the graph, like green or yellow cars.

2

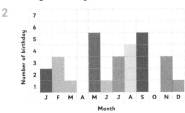

Month

Challenge

1   Teacher to check.

2   a   Teacher to check. For example, it's about which is the most/least popular season.

   b   Teacher to check. For example, that summer got the most with 8 and autumn the least with 4.

   c   Favourite seasons.

Mastery

1   Teacher to check. One possible answer is:

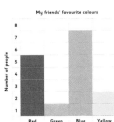

My friends' favourite colours

2   Teacher to check. One possible answer is:

Answer	Question
8 people	How many people had 11 lollies?
9 lollies	What is the least amount anyone had?
9 and 13+ lollies	Which 2 amounts had the same?
7 people	How many people had 12 lollies or more?
20 people	How many people completed this task?

UNIT 9: Topic 1

Practice

1   impossible, certain, maybe, maybe

2   a–b   Teacher to check.

Challenge

1

Bag of marbles	Chance
	maybe
	certain
	impossible
	maybe

2   Teacher to check. One possible answer is:

Description	Marbles
It is impossible to pick a yellow marble.	For example, 4 blue marbles.
It is certain you will pick a blue marble.	For example, 4 blue marbles.
You will maybe pick a green marble.	For example, 1 green, 2 blue, 2 yellow marbles.
Red has the most chance of being picked.	For example, 4 red, 2 yellow marbles.

Mastery

1   a–d   Teacher to check.

2   a–d   Teacher to check.

OXFORD UNIVERSITY PRESS